VERY SHORT INTRODUCTIONS · FOR CURIOUS YOUNG MINDS

W0044187

The
SHOCKING
Truth about
ENERGY

Dr Mike Goldsmith

OXFORD
UNIVERSITY PRESS

OXFORD
UNIVERSITY PRESS

Great Clarendon Street, Oxford OX2 6DP

Oxford is a registered trade mark of
Oxford University Press in the UK and in certain other countries

© Oxford University Press 2023
Text written by Dr Mike Goldsmith
Illustrated by Ekaterina Gorelova and Ana Seixas

Designed and edited by Raspberry Books Ltd

The moral rights of the author and artist have been asserted
Database right Oxford University Press (maker)

First published 2023

All rights reserved.

British Library Cataloguing in Publication Data:

ISBN 978-0-19-278291-5

1 3 5 7 9 10 8 6 4 2

Printed in China

Paper used in the production of this book is a natural,
recyclable product made from wood grown in sustainable forests.
The manufacturing process conforms to the environmental regulations
of the country of origin.

Acknowledgements

The publisher and authors would like to thank the following for permission to use photographs and other copyright material:

Cover artwork: Ekaterina Gorelova and Ana Seixas; Photos: Pavlo S/Shutterstock; Aleksandr Bryliaev/Shutterstock and author.
Inside artwork: p1: Pavlo S/Shutterstock p9: Mark Rademaker/Shutterstock; p9(br): Georgios Kollidas/Shutterstock; p14: Anita van den Broek/Shutterstock; p16: flower travelin man/Shutterstock; p28: Nicku/Shutterstock; p31: posteriori/Shutterstock; p36: macro videography/Shutterstock; p59: mikroman6/Moment/Getty Images; p68: Kim Steele/Photodisc/Getty Images; p70: Everett

Collection/Shutterstock; p72: Stocktrek Images/Getty Images; p73(t): muratart/Shutterstock; p73(b): Heritage Images/Hulton Archive/Getty Images; p80: Morphart Creation/Shutterstock; p86: IfH/Shutterstock; p87: BlueOrange Studio/Shutterstock.

Artwork by **Ekaterina Gorelova**, **Ana Seixas**, Aaron Cushley, Oxford University Press, and Raspberry Books.

Every effort has been made to contact copyright holders of material reproduced in this book. Any omissions will be rectified in subsequent printings if notice is given to the publisher.

Images are to be used only within the context of the pages in this book

Did you know that we also publish Oxford's bestselling
and award-winning **Very Short Introductions** series?
These are perfect for adults and students
www.veryshortintroductions.com

Contents

Chapter 1

What is Energy?

Science is all about simplifying the way the world works. One very simple but very scientific way to understand it is to say that the world is made of just two things: matter and energy.

Matter includes most of the things you can see around you: solids like the earth, liquids like water, and gases like air. Energy is everything else, like light, heat, sound, and electricity.

Speak like a scientist

ENERGY

The word 'energy' comes from a word from ancient Greece, 'energia', which means 'activity'. Energy is what you have to have in order to do work. When you whack a ball, you give it some energy.

Energy exists in many forms. Imagine a thunderstorm—as well as the light energy and sound energy of lightning and thunder, the falling raindrops and rushing wind in a storm have energy too: the energy of motion.

What happens when lightning strikes

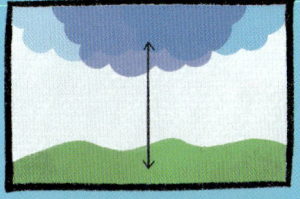

1 Invisible electrical energy moves between clouds and the ground.

2 Light energy, other forms of radiant energy, and heat energy are released.

3 Air molecules suddenly gain motion energy and pass it on to their neighbours . . .

4 . . . resulting in sound energy.

When anything happens, energy changes are involved. When electrical energy flows from the clouds as lightning, some of that energy changes to heat energy in the atmosphere. Other parts of it change to the light and sound energy we see and hear.

Every machine and gadget ever made works by changing energy from one form (or forms) to another. And our bodies are energy-changing machines too, as we'll discover on page 14.

Potential energy

Blowing a ping-pong ball off a table's edge so that it falls is easier than blowing it along the table for the same distance, because in the first case, your breath only has to start it moving—once off the table it falls naturally. You could say it has the **potential** to fall. The higher above the ground something rises, the more potential energy it gains, and that's why you would need **a lot more puff** to get the ball to go up a slope—you are giving it more potential energy as it rises.

An apple on a tree has potential energy. When it falls, some of its potential energy changes to kinetic energy as it's falling, then to a little thermal energy (heat) when it stops.

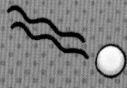

WHAT HAPPENS WHEN AN APPLE FALLS?

POTENTIAL ENERGY
(when attached to the tree)

KINETIC ENERGY
(when falling off the tree)

THERMAL ENERGY
(when it hits the ground)

Potential energy isn't always related to height. If you stretch a rubber band between your hands, the energy your muscles supply to it is stored in the rubber. When you let go, some of the potential energy within it changes to kinetic energy. If it snaps, some of the potential energy has also turned into sound.

Speak like a scientist

FORCE

Force is a push or pull. If you apply a force to an object, that object will change in some way. If the object is a ping-pong ball flying towards you, it will change direction if you whack it with a bat or it whacks you in the head.

WORK

Work is what a force does to an object. When you whack that ball, you change the direction it's moving in. That change shows that you have done work on it.

POWER

Power is related to energy, but it involves time. Whether you run or walk upstairs, the same amount of energy is involved, but running takes more power because it happens more quickly.

PRESSURE

Pressure is force applied over an area. If you stand on mud, wearing ordinary shoes, you will leave shallow footprints behind. But if you wear studded footwear, like football boots, the prints of the studs will be deeper. As your weight is the same in both cases, the downward force must be the same too, but the area in contact with the ground is much smaller with the studs. And the same force applied over a smaller area means a higher pressure and, therefore, deeper holes.

ENERGY HERO

ISAAC NEWTON

Worked out the laws that moving and falling things obey. Also untangled some of the mysteries of light, and invented new kinds of mathematics and a new sort of telescope.

Simple machines

Nearly 2,500 years ago, a genius called Archimedes did something that his strongest friends couldn't do (and the Greeks were a sporty lot—they even invented the Olympics!). He single-handedly dragged a loaded ship into harbour using a compound pulley he had invented.

The compound pulley

The compound pulley, like all simple machines, works by changing the force that the user applies in some way. Its secret is that by pulling one rope a long distance fairly gently, another rope moves a far shorter distance, but much **more forcefully.**

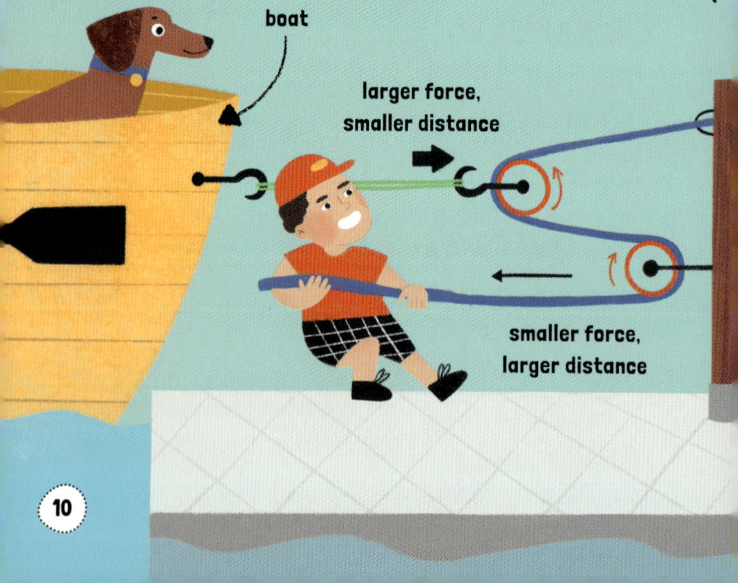

boat

larger force,
smaller distance

smaller force,
larger distance

ENERGY HERO

ARCHIMEDES

Ancient Greek thinker
who developed new kinds
of mathematics, and
invented things too.

This is a very short introduction to energy and
the shocking truth about it. You'll discover that . . .

our **thoughts**
are made of
electrical energy

a bag of sugar has the
same energy as a **trillion
nuclear bombs**

fridges are hot . . .
and coldness
doesn't exist

a black sandwich
can become incredibly
dangerous

Chapter 2

Kinds of Energy

Energy comes in many forms. These are the main kinds:

chemical

electrical/ magnetic

potential

nuclear

kinetic

TYPES OF ENERGY

radiant (including light)

mass

sound

thermal (heat)

As we've already seen, energy can change from one form to another, and machines are inventions which make this happen. Many machines deal with lots of kinds of energy at once, but one of these is almost always **electricity**. A phone takes in sound when you speak to it, light when it takes a photo, and radio waves when it picks up a signal.

All of these change to electricity inside the phone, and more electricity is supplied from chemical energy in the battery.

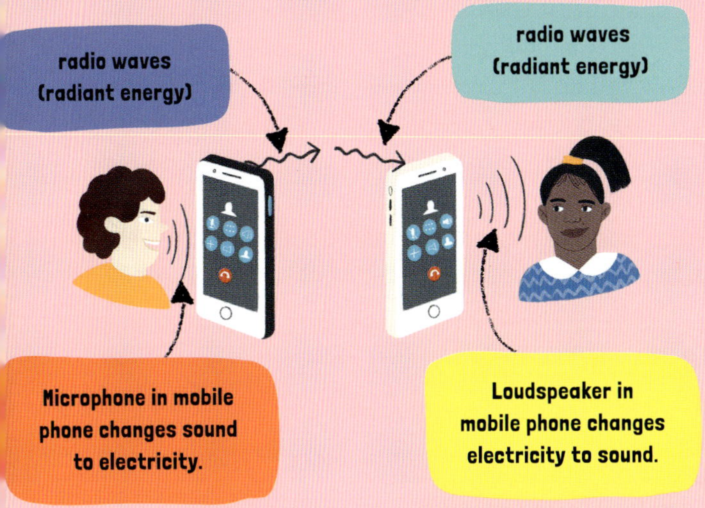

radio waves (radiant energy)

radio waves (radiant energy)

Microphone in mobile phone changes sound to electricity.

Loudspeaker in mobile phone changes electricity to sound.

Your phone also changes electricity into sound when you listen to it, into light when the screen glows, and into radio waves when it sends a signal. If you drop it, the potential energy it has due to its height above the ground changes to kinetic energy when it falls, and then to heat and sound when it hits the ground.

All forms of energy except nuclear have been known and studied for a long time. But the idea that motion, light, and the rest are **all different forms** of the same thing is only about two centuries old.

Living machines

Our bodies use most kinds of energy and constantly change it from one form to another. As soon as we eat food, our bodies begin to break it down, changing its energy-rich chemicals into other chemicals which can be dissolved in blood. The heart **pumps the blood** around the body, and some of the chemicals are broken down in the muscles, which convert the chemical energy into kinetic (motion) energy. Other chemicals are stored as fat; locking up the energy until we need it.

All these changes release heat (thermal energy), which flows constantly out of our bodies, carried away by our breath and by a type of radiant energy called **infrared** from the skin. Infrared can be seen by heat-sensitive cameras.

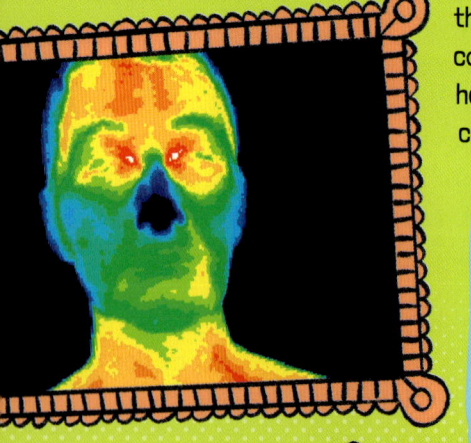

Pictures from these cameras show the hottest areas as white and red, and the coolest areas as green and blue.

MAGNIFIED CROSS-SECTION OF SKIN

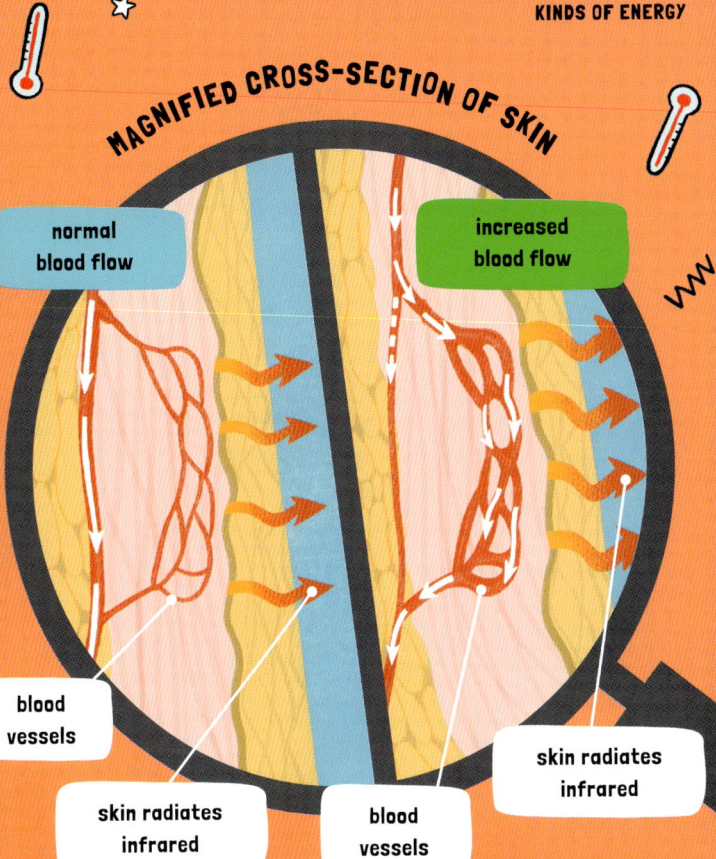

normal blood flow

increased blood flow

blood vessels

skin radiates infrared

blood vessels

skin radiates infrared

To shed heat, your body increases blood flow to small blood vessels (tubes) just under the skin. The skin radiates more infrared.

In winter, we sometimes need to eat extra food just to keep ourselves warm. Whether the food itself is hot makes little difference—it is the chemical changes that **increase our temperature.**

Thermal energy (heat)

Thermal energy (heat) is **almost always** produced whenever any other energy change occurs. When you run, chemical energy changes to kinetic energy, but also to thermal energy, which is why you feel hot. When phones, laptops, and TVs change electrical energy to sound and light, they also warm up a bit.

Thermal energy always **s p r e a d s** until it has evened out. Another way to say this is that hot things always cool down, given time.

How thermal energy spreads

Any series of energy changes will end with heat, and once that heat is evenly spread, no further energy changes are possible.

 shiver

 Brr

The Big Chill

This rule applies to the entire Universe. The whole place is cooling, and many scientists believe it will go on doing so for ever. This is called the Heat Death Theory, or sometimes, The Big Chill. Eventually, the Universe will be the same very low temperature all over, and nothing more will happen—ever.

Did someone turn the heating off?

✳ Speak like a scientist ✳

THERMOCOUPLE

Until heat has evened itself out, it can be turned into other kinds of energy. An invention called a thermocouple changes heat to electricity, but it only works if one end is colder than the other. Electricity is produced when heat flows along it from the hot end to the cold one. If the heat is evened out so that both ends are the same temperature, there's no heat flow, and no electricity.

Kinetic energy (motion)

In space, flecks of paint or metal from long-gone rockets and crumbly old satellites are whizzing round the Earth at speeds of up to 8,000 metres per second, about ten times faster than a bullet. At those speeds, even such tiny things could **wreck a satellite** or even kill an astronaut.

There are two parts to the kinetic energy of an object—its **speed** and its **mass** (mass is closely related to weight—see page 36). But the speed counts more: doubling the weight of a moving object doubles its kinetic energy, but doubling its speed multiplies its energy by four. That's why car crashes are **much** more than **twice as dangerous** if they happen at 60 kph than at 30 kph.

Things don't need to move from place to place to have kinetic energy. A spinning top, coin, or ballerina has it, and so does anything that wobbles —from a plucked guitar string to **a plate of jelly.**

Sound energy

Sound is a kind of motion—a wobble that can travel through any kind of matter. While the sound-wobbles (called waves) are constantly moving onwards, the wobbling things (usually tiny particles called molecules) just **shake** backwards and forwards a little.

Wave patterns

Here are four snapshots of a sound wave, a millionth of a second apart. The molecules each wobble backwards and forwards, while the whole wave-pattern moves from left to right.

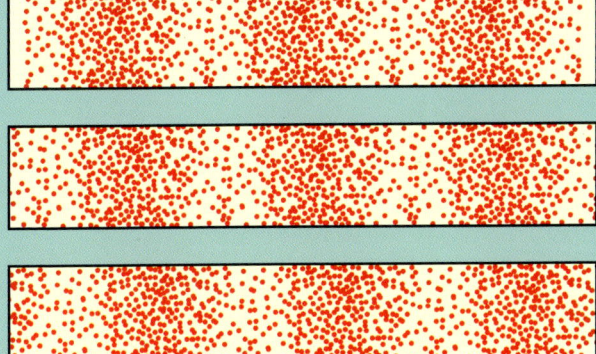

The rate at which the molecules wobble backwards and forwards is what we hear as pitch: rapid wobbles sound high, like whistles; sluggish ones sound low, like distant **rumbling thunder**. We humans cannot hear very high or very low waves, but elephants can hear very low-pitched sounds from underground through their feet, and bats hear very high-pitched sounds (ultrasound) and are so good at it they can use the sounds to 'see' in the dark. Sometimes sounds can be heard over huge distances: some whales can hear other whales on the far side of the Earth.

Light and other forms of radiant energy

Light is always moving—very fast indeed. It's fast enough to go round the world in under a second. It can travel through most gases, some liquids, and a very few solids, like window glass. We call the things it can travel through 'transparent'. Unlike sound, though, light can travel through space, which is why we can see the Sun but not hear it.

Mum, is that you?

Waves, photons, and radiation

Light is made of tiny particles of energy called photons, but it behaves more as if it is made of waves. These are not very different to waves on the sea, with many different lengths and strengths.

0.000000001 mm	0.0000001 mm	0.00001 mm	0.001 mm	0.1mm	1cm	1m	100m
gamma rays	X-rays	ultra violet	infrared			radio waves	

light (which is the radiation we see)

Red light waves are longer than green ones, and green light waves are longer than violet. Some animals, like snakes and bees, can see waves longer than red (these are called infrared) or shorter than violet (**ultraviolet**). There are waves longer or shorter even than these, which no living thing can see but which we can measure. These include gamma rays, which come from rare and dangerous materials like **uranium**, X-rays (used in hospitals to see inside people), and radio waves (used by mobile phones). All these types, including light, are called radiant energy, or **radiation**.

Types of radiant energy

Light can carry a lot of energy. In a laser, light is sent in a tight beam. When it shines on an object, the light changes to thermal energy. Powerful lasers can use this effect to melt metal.

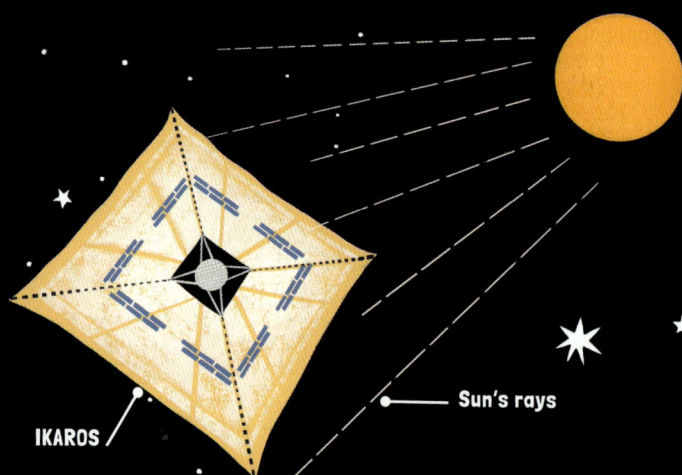

IKAROS

Sun's rays

Light can also push. Scientists have built solar sails to catch the light from the Sun and use it to push them around the Solar System. The first one that worked was called IKAROS and used a thin plastic sail coated in shiny metal. Launched in 2010, it sailed through space for many years.

Energy machines

Today, we are surrounded by machines, such as vehicles and gadgets, and most of them use electrical energy. Vehicles usually carry fuel on board, such as petrol (though more now run on electricity). They burn this fuel to get the energy they need to move. Many smaller machines—gadgets, like mobile phones—use batteries to supply them with electrical energy, while others, like washing machines, televisions, and kettles, plug into a household electricity supply.

Nearly all the electricity used by machines comes from power stations which burn fuel (often coal, oil, or gas) to generate electricity. The wind, tides, waves, falling water, the hot depths of the Earth, and nuclear materials (to be investigated with caution in Chapter 5) also supply some machines with energy—almost always in the form of electricity.

An energy storage system

Energy created from wind, water, or sunlight is a challenge because it might not be needed immediately. For example, in the case of sunlight, we might want to store the energy collected on a **hot** sunny day to warm us on a cold dark night. Batteries store energy, but they are heavy, expensive, wear out, and waste a lot of the energy fed to them. Instead, we can convert the electricity we make from sunlight to potential energy, by using the electricity to pump water from lower to higher.

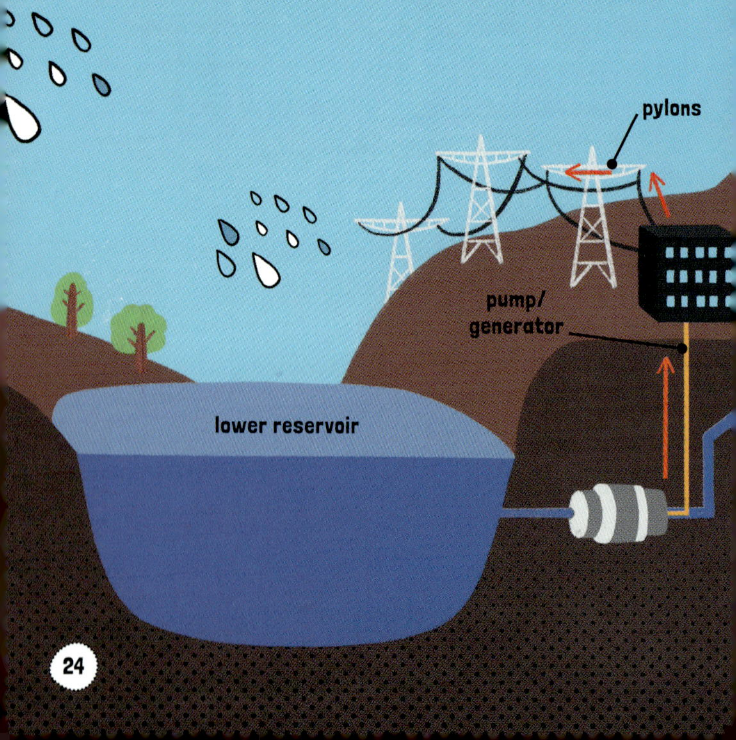

pylons

pump/
generator

lower reservoir

This stores potential energy which can be turned back to electricity by allowing the water to **flow** out of the higher reservoir again. The kinetic energy of this moving water can be turned into electrical energy by using a **dynamo** (see page 62).

electricity cable

upper reservoir

Water is pumped up to store spare electrical energy as potential energy . . .

. . . and allowed to flow down again, to turn the potential energy back to electrical energy when needed.

Electrical and magnetic energy

Electrical energy is transferred from power stations to homes and businesses through cables. And, through similar cables and wires, we transmit information too—the internet is mostly a network of electrical connections.

Our bodies use electrical energy as well—**it's what our thoughts are made of** and it's the way we tell our muscles what to do. Our eyes, ears, and skin send our brains electrical messages too. We make this electrical energy from the chemical energy in food.

brain

nervous system

Speak like a scientist

ELECTRONS, ATOMS, AND MOLECULES

Electricity is all to do with tiny particles called **electrons**. They prefer to live in **atoms** and molecules, but they can easily be persuaded to leave. This sometimes happens when you take your sweater off—the crackles and glows are sound and light energy released by electrons moving around. We see those flashes and hear those crackles because our eyes and ears send electrical signals to our brains.

It is electrical energy that holds **atoms** together. Electrons make up the outer parts of an atom, and each electron has an electrical charge. The centre of an atom, the **nucleus**, also has an electrical charge, but it is opposite to that of the electrons. These two different kinds of charge **pull** on each other, and this pull holds the atom together. (On the other hand, two electrons, or two nuclei, will repel each other—push each other away—because they have the same type of charge.)

How atoms make a molecule

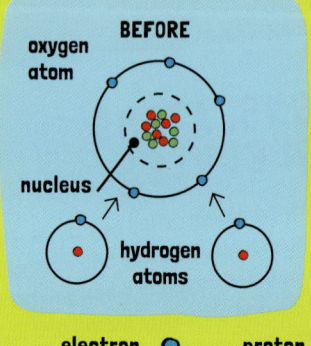

BEFORE

oxygen atom

nucleus

hydrogen atoms

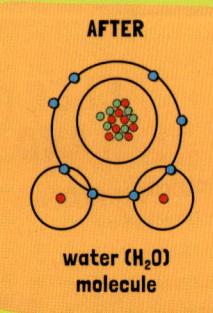

AFTER

water (H₂O) molecule

electron 🔵 proton 🔴 neutron 🟢

Electrical energy holds **molecules** together too. When two or more nuclei (more than one nucleus) pair up to form a molecule, some of their electrons are shared, and again, it's the pull between the electrons and nuclei that hold the molecule together.

Magnets and electricity are closely related. In fact, whenever electricity flows down a wire, the wire turns into a magnet (though usually a very feeble one).

ENERGY HERO

MICHAEL FARADAY

Invented the electric motor and the dynamo (generator). Also, a chemistry genius.

In 1821, Michael Faraday discovered how to use a magnet to change electrical energy into kinetic energy (energy of motion), when he made a wire dance around a magnet.

The world's first electric motor

Faraday's first **motor** only demonstrated how electrical energy could be changed into kinetic energy, but soon after, practical versions were built. The first motor was used to print books.

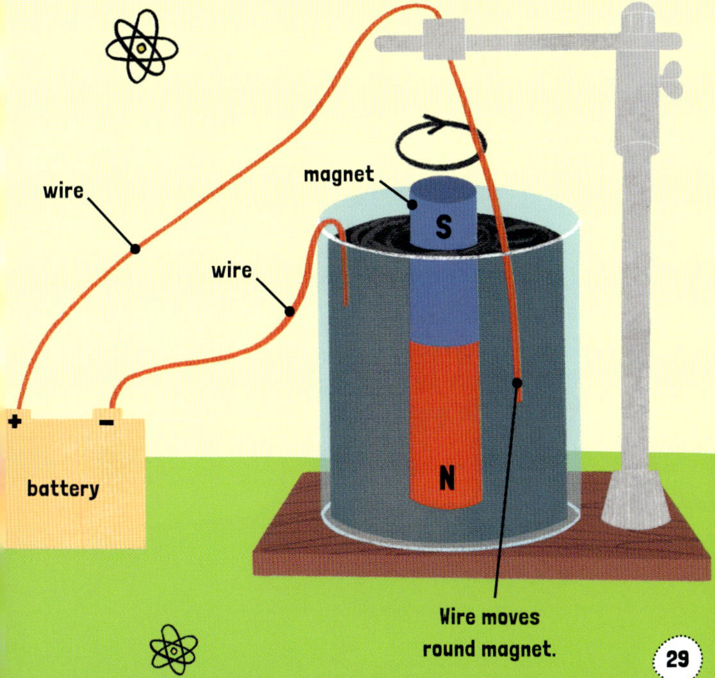

wire

magnet

S

wire

+ −

battery

N

Wire moves round magnet.

Chemical energy

Humans have been making **explosions** for over a thousand years, and for most of that time we've used a chemical mixture called gunpowder.

Like other explosives, gunpowder works by quickly converting the energy it contains (which is called chemical energy) into light, heat, and sound. In other cases, chemical energy is released more slowly—in your intestines, for instance, where chemicals from your body break down the chemicals in your food in order to digest your breakfast. Sometimes chemical energy can take centuries to release, as when metals slowly rust away.

Speak like a scientist

CHEMICAL REACTIONS

A chemical reaction is what happens when atoms are swapped around between different molecules. For instance, molecules of the poisonous green gas chlorine will happily combine with those of the soft metal sodium to produce sodium chloride (salt).

When salt is produced, chlorine atoms start off joined to other chlorine atoms in chlorine molecules, but end up being joined to sodium atoms in sodium chloride. When this happens, lots of thermal and light energy is released.

How salt is made

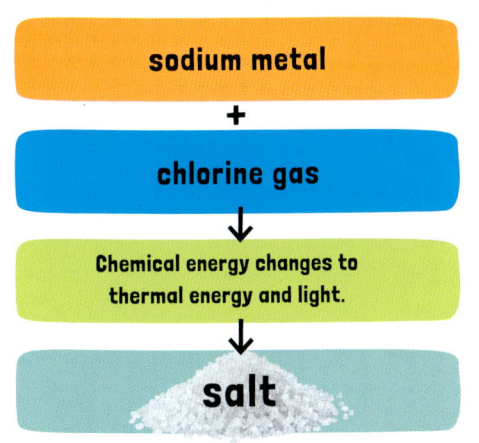

sodium metal

+

chlorine gas

Chemical energy changes to thermal energy and light.

salt

Chemical reactions need something to start them off. For instance, paper contains chemicals which will react with **oxygen** molecules from the air to produce thermal energy, radiant energy, and molecules of **carbon dioxide** and water. In other words, paper will burn. But this won't happen until enough thermal energy is available to start breaking apart the molecules in the paper and the oxygen molecules from the air. This is why paper will only burn if you heat it up enough.

The reason that thermal energy is needed to start many reactions is that the molecules will only start swapping atoms around if they collide at a high enough speed—that is, they need plenty of kinetic energy. And the way to increase the kinetic energy of molecules is to add heat (thermal energy) to them. In most cases, once the reaction has begun, it will release thermal energy of its own, and this will keep the reaction going.

Why is extra kinetic energy needed? It makes molecules of different kinds slam into each other so they break up into atoms, and these atoms stick to other atoms to form new molecules.

Oi! Watch where you're going

Nuclear energy

Like chemical energy, nuclear energy is stored in atoms, but it hides much deeper inside and is usually much harder to get out. But when it does escape, there's a lot more of it: an atom contains **millions of times** as much nuclear energy as chemical energy.

✳ Speak like a scientist ✳

THE NUCLEUS AND THE STRONG NUCLEAR FORCE

The nucleus of an atom is made of joined-together particles called neutrons and protons. To release the nuclear energy within the nucleus, these particles must be pulled away from each other. You might think this would be easy, because protons repel each other.

So, a strong force must be at work, overcoming this repulsive force and thereby holding the protons close together. This is called the strong nuclear force.

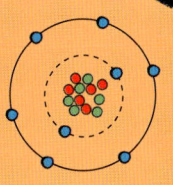

Electrical charges come in two types: positive and negative. Electrons are negatively charged; protons are positively charged. Negatively charged particles repel each other, but attract positively charged ones, so electrons repel other electrons, but attract protons. Meanwhile, positively charged particles repel each other, so protons repel protons. A quick way of saying this is: unlike charges attract; like charges repel.

Magnets work in just the same way: unlike poles attract; like poles repel. So, a magnet's north pole will attract the south poles of other magnets, but repel north poles.

If nuclei contain enough radioactivity protons, their repelling force is a match for the strong nuclear force, so the nuclei will fall apart all by themselves. When they do this, a mixture of small, high-speed particles and high-energy gamma radiation is released too. The mixture is called **radioactivity** (see page 65).

The protons in the nucleus of an atom attract the electrons that surround the nucleus (this is what holds the electrons in place).

Mass

The difference between a heavy thing and a light one is mass: the **heavy thing** has more mass. When we weigh something, we are measuring its mass. But mass and weight aren't exactly the same. When people and things are in orbit around the Earth, they have no weight but they still have mass. Weight is actually a force—the force that a mass feels due to the gravity of the Earth or of another body, like the Sun or the Moon. So, what is mass?

SUGAR

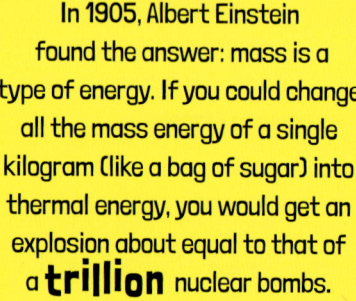

In 1905, Albert Einstein found the answer: mass is a type of energy. If you could change all the mass energy of a single kilogram (like a bag of sugar) into thermal energy, you would get an explosion about equal to that of a **trillion** nuclear bombs.

ENERGY HERO

ALBERT EINSTEIN

Discovered that mass is energy, what gravity really is, how time is affected by motion and by gravity, and helped prove that light and other forms of radiant energy are particles. Also helped prove that atoms exist.

Luckily, mass doesn't change to other forms of energy quite as easily as the other types. One way to do it is to find or make a substance called **antimatter** and allow it to touch ordinary matter. The enormous explosion you get is mass energy changing to kinetic, radiant and thermal energy, then sound.

Speak like a scientist

ANTIMATTER

Matter is made of particles, like electrons. Antimatter is made of similar particles, but the particles are slightly different. For instance, an electron is negatively charged, but an antimatter electron (called a positron) is positive (hence its name).

Energy rules the world—but what rules does it obey?

The next chapter has **all the answers.**

Chapter 3

How Energy Works

Like every other bit of science, energy can't be studied or used properly unless it can be measured, and to do that, someone needs to decide how much energy counts as a single unit.

There are a few units of energy in use today, but by far the most popular is the **joule**—the jewel in the crown of energy units. It takes about a joule of energy to lift an apple from the floor to a tabletop.

The **power** of a machine is usually measured in **watts**. To do something more quickly needs more power. Let's say that lifting the apple took a second. That would mean the power involved was one watt. If you raised it to the same height in one-tenth of a second, it would take 10 watts.

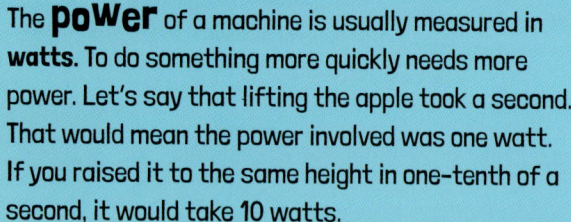

The energy our bodies can obtain from food is measured in calories. There are 4,184 joules in a calorie.

Armed with these handy **units**, we can compare all sorts of things (all these are rough figures):

Mass of a ping-pong ball:
250 trillion joules

Electrical energy of a strike of lightning:
1,000 million joules

Chemical energy in a biscuit:
250,000 joules (60 calories)

Thermal energy in a cup of tea:
50,000 joules

Kinetic energy of Usain Bolt sprinting:
7,000 joules

Electrical energy of a thought:
100 joules

Power of the Sun:
400 million, million, million, million watts

Power of a kettle:
2,400 watts (often written as 2.4 kW,
because 1,000 watts is one kilowatt)

Brightness of a star:
one millionth of a millionth of a watt

Power of a whisper one metre away on your eardrums:
one hundredth of a millionth of a millionth of a watt

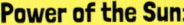

The first law of thermodynamics

The most important scientific discovery about energy is that it is conserved, which means it cannot be created nor destroyed. All it can do is change from one form to another. And, when one form does change to another, the total amount of energy is unchanged: there is just the same amount of energy in the apple on page 7 whether it is in the form of potential energy, kinetic energy, or thermal energy.

Many people contributed to this discovery, but one of the most important was Gabrielle Émilie Le Tonnelier de Breteuil, Marquise du Châtelet—usually known as Émilie du Châtelet. She began by studying kinetic energy, and this is still one of the easiest ways to understand energy conservation.

✳ Speak like a scientist ✳

THERMODYNAMICS

The scientific study of thermal energy, and how it changes to and from other forms of energy.

As a pendulum rushes from side to side, it slows down as it swings up, and speeds up again as it swings down. This means its kinetic energy keeps growing and falling—but how can it do this? The answer is that its potential energy changes all the time, growing when the swing moves up, falling when it moves down. But the total energy is conserved—that is, it stays the same throughout the up and down movement. If the pendulum has ten joules of energy at the top of the swing, it has ten joules (J) at the bottom of the swing too: those ten joules have been conserved. It's just that the energy shifts from kinetic to potential and back.

Energy of a pendulum

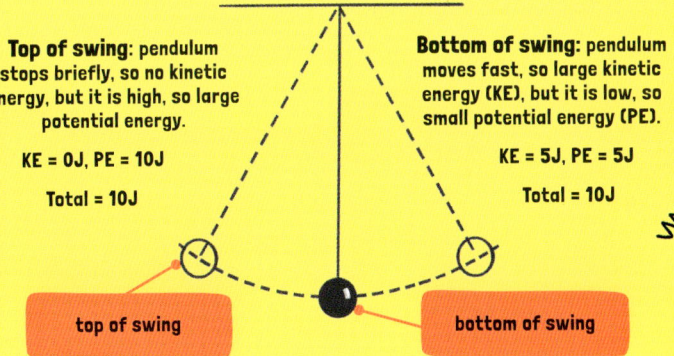

Top of swing: pendulum stops briefly, so no kinetic energy, but it is high, so large potential energy.

KE = 0J, PE = 10J

Total = 10J

Bottom of swing: pendulum moves fast, so large kinetic energy (KE), but it is low, so small potential energy (PE).

KE = 5J, PE = 5J

Total = 10J

top of swing

bottom of swing

When eventually the pendulum slows down and stops, there's still the same amount of energy—but all the kinetic energy is now changed to thermal energy.

ENERGY HERO

ÉMILIE DU CHÂTELET

Worked out exactly what kinetic energy is and that all energy must be conserved.

The question of why energy is conserved was answered by Emmy Noether. Her theory is complex, but she managed to prove that if energy was not conserved, a pendulum might swing differently on different days, or the mass of an object might be higher from one year to another. In other words, she discovered that the conservation law is built into the way the world works.

ENERGY HERO

EMMY NOETHER

Worked out why it is that energy is conserved. Helped develop one of the most important types of mathematics, called abstract algebra.

Using the first law of thermodynamics

The law of conservation of energy is brilliant. Not only does it tell us something amazing and unexpected about the whole Universe, it also means we can find out about things that we can't easily measure, like chemical or nuclear energy.

The law means that if we can convert those types of energy to easily measured types, like thermal or electrical energy, we can measure those instead. For instance, to measure the **chemical energy in a biscuit,** we can burn one and measure the thermal energy produced. Thanks to the conservation law, you can be confident that this same amount of energy is stored in another biscuit for your body to use if you eat it.

The conservation law also tells engineers how much energy is lost by a vehicle. That energy loss must be paid for with extra fuel. Reducing the loss (also called 'increasing the **efficiency**') will save fuel.

When energy is lost from a machine, some will appear as sound energy—the more **screechy and groany** it is, the more it is losing. But most lost energy is in the form of heat, and the less efficient the machine is, the hotter it and its surroundings will get. In fact, once the graunching noises have been heard, they fade away too—and in doing so, they add a tiny amount of heat to the air, and their listeners' ears.

✳ Speak like a scientist ✳

GRAUNCHING

A crunching, grinding noise, made by unhappy machinery.

Oh, stop graunching and get on with it.

Is heat the same as temperature? (Spoiler alert: no)

In some ways, heat (thermal energy) is the easiest form of energy to understand because we can feel it, and it appears in every aspect of our lives: from frozen puddles to boiling kettles, and from hot hamburgers to cold icebergs. Heat is simply the kinetic energy of molecules (or other similar things like atoms). If you touch something made of fast-moving molecules, the molecules in your skin will start moving faster too—that is, your skin will warm up.

The relationship between heat and temperature can be confusing. The temperature of something depends on how fast its molecules are moving: the faster they go, the higher the temperature. Hot water has a higher temperature than cold water, because its molecules are *moving faster.*

A giant iceberg contains a lot of heat—the bigger the iceberg, the more it has. Think about all those particles wobbling—trillions and trillions of them. Even when molecules can't move from one place to another, such as when they are in a block of ice, they can still have lots of kinetic energy, because they can vibrate.

There are millions of times more molecules in an iceberg than in a cup of coffee, so the iceberg contains a lot more heat than the coffee does. But because the iceberg's molecules are slower moving than those in the tea, it has a lower temperature.

Everyone had been confusing heat and temperature more or less for ever, until in 1762 Joseph Black showed that adding heat to melted ice does not increase its temperature, because the heat energy goes into pulling the molecules away from one another.

From ice to cold water

-2°C	0°C	0°C	0°C	2°C
ICE	ICE	SLUSH	WATER	WATER

ENERGY HERO

JOSEPH BLACK

Discovered the difference between heat (which is energy) and temperature (which isn't). Also helped us understand carbon dioxide gas, and magnetism.

In the same way, when we put a pan of water on a gas ring, the thermal energy flows from the burning gas to the water. Once the pan of water is boiling, it doesn't get any higher in temperature: it remains at 100°C. But the fact that it is still sitting on the burning gas ring shows that thermal energy is still flowing in. This energy is now being used to change the water to the steam and vapour that bubbles from it.

Because temperature is different to heat, it is measured in different units: degrees Celsius (°C) usually, though other units are available. Heat, like most forms of energy, is measured in joules.

Thermal energy is kinetic energy

We know that energy cannot be destroyed (because it is conserved—see page 40) and that it ends up as heat (see page 16). And we now have our joules—units to count it in. This gives us a way of measuring the energy of **anything we like**. Or indeed, dislike.

The chemical energy in food, for instance, is measured by burning a sample of food in a device called a bomb calorimeter which measures the heat energy produced.

Kinetic energy can be converted to heat to be measured too: in fact, that's how energy scientist James Joule worked out the size of his energy unit in the first place.

ENERGY HERO

JAMES JOULE

Found out how thermal and kinetic energy are related, and studied the relationship between heat, magnetism, and electricity.

JOULE'S ENERGY EXPERIMENT

> I dropped a weight on a string to spin a paddle wheel.

The wheel was underwater, and he measured the increase in temperature of the water.

weight

paddle wheel

The weight of the falling weight and the distance it fell gave him the kinetic energy of the wheel (A).

The temperature rise, multiplied by the weight of the water, gave him the heat energy (B).

And the conservation law told him that these energies must be the same (A=B).

Joule found that one kilogram falling 10 cm is about one joule of energy. And 4,200 joules of energy raises the temperature of one kilogram of water by 1°C. To make a cup of tea this way, you'd have to drop a weight of more than one hundred tons.

The second law of thermodynamics

The 'hot things cool' law (page 16) rules the Universe: a thing that is hotter than its surroundings will cool. It will cool slowly if the thing is a baked potato, or quickly if the thing is a hot pan plunged into cold water. On the other hand, once you take an ice cream out of a **freezer,** it will gain heat. This time the 'hot things' are the air and your hand, which get a bit cooler when they transfer just enough heat to the ice cream to melt it. But why does this happen?

We can find the answer if we think about heat as motion. When a fast (= hot) particle (let's call it Paulette) bangs into slower (= cooler) ones like Pedro, Patricia, and Pasha then Paulette will slow down and cool down. But the slow particles will speed up and **heat up.**

slower (= cooler) particles

Paulette
Patricia
Pedro
Pasha

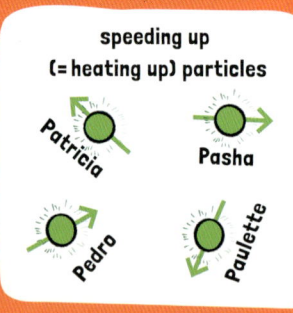

speeding up (= heating up) particles

Patricia
Pasha
Pedro
Paulette

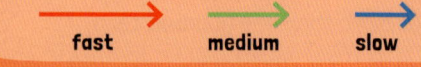

fast   medium   slow

Why heat evens out

In other words, some of the Paulette's kinetic energy has been shared with the slower particles. They will go on to speed up many more particles a little, until all the particles are travelling at about the same speed. Once this has happened, the heat has evened out.

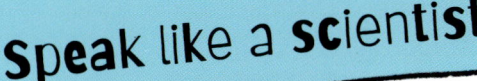

Speak like a scientist

THE SECOND LAW OF THERMODYNAMICS

'Hot things cool'—or 'heat spreads until it's evened out'—is the second law of thermodynamics.

Fridges are hot!

If everything that happens makes heat, how do fridges make cold? **They don't.**

A fridge is a heat-moving and heat-making machine, not a heat-destroying machine (that would be impossible). It moves heat out of itself to keep your milk fresh, but that heat warms up the kitchen.

Fridges

The amount of heat that comes out of the back of a fridge is more than the amount it takes from its interior. So, every fridge is a heater in disguise—the world gets a little warmer every second one is working.

A fridge can't make coldness because **coldness doesn't exist**; there's just heat, or the lack of it. Think about darkness, silence, and stillness—they aren't things in themselves, but rather the lack of light, sound, or motion.

How a fridge works

1 It uses electricity to power . . .

2 a compressor, which squeezes a gas called a refrigerant until . . .

3 it turns into a liquid. The refrigerant molecules have to slow down to do this, which means they lose kinetic energy.

4 This kinetic energy changes to thermal energy, which heats the kitchen.

5 The liquid refrigerant then turns back into gas inside pipes inside the fridge. (The pipes are usually hidden away behind a panel at the upper part of the back of the fridge.) Changing to a gas means the refrigerant molecules speed up again, gaining kinetic energy, which comes from thermal energy inside the fridge. In other words, the refrigerant takes heat from the inside of the fridge as it changes from liquid to gas (evaporates).

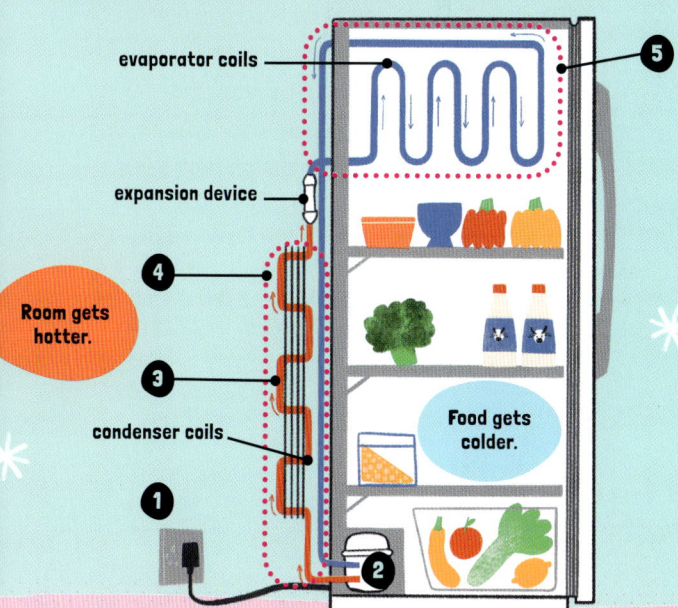

evaporator coils — **5**

expansion device

4

Room gets hotter.

3

condenser coils

1

Food gets colder.

2

Humans took control of the world because they found out **how to use energy. The next chapter tells the story.**

Chapter 4

Powering the Planet

Humans have been converting chemical energy to thermal energy by setting fire to things (wood, mostly) for over a million years. This thermal energy was used to keep warm and to cook mammoth burgers, while the light energy from the fire kept dangerous animals away, or at least make it more difficult for them to creep up on people at night. And we've been burning things ever since.

The energy we use to light and heat our homes and cook our burgers still comes mainly from burning fuels—either gas in central heating systems and cookers, or coal or oil to produce electricity in power stations.

Today, about 84% of the world's power comes from power stations which burn **fossil fuels**, like coal, which is formed from the remains of plants that died millions of years ago. These plants grew in vast swamps, covering millions of square kilometres. When they died, they were first covered by water and then by other dead plants, which changed slowly to soil and then to rock. Over millions of years, the pressure from the layers of rock above and the heat from below (see page 87) crushed and changed them until they became coal.

✷ Speak like a scientist ✷

EFFICIENCY

The percentage of useful kinetic energy produced by a machine is called its 'efficiency'.

James Watt invented a very popular steam engine in 1776, though it was only 4% efficient. In those days, scientists hadn't got round to defining efficiency properly, so no one could be sure what was what.

After Watt's invention caught on, there were all sorts of steam-powered factories and steam trains too. Factories could now be supplied with raw materials by train and deliver their products the same way.

Ships could be steam-powered as well. At last, sailors, traders, and travellers didn't have to wait for winds to blow their sailing ships wherever they wanted to go. Now that crossing the sea was faster and more reliable, trade between distant countries sped up. Before long, many other countries joined in this exciting new **Industrial Revolution.** More factories sprung up, trade increased, and with it, the world's population. It was the start of our modern world: full of technology, travel, and shopping.

The Industrial Revolution has not been all good. Many people who moved from the countryside to cities in search of work found their lives were worse than before. Industrial pollution, mainly from burning fossil fuels, damaged the health of people and animals. And, as we'll soon discover, it's given us a warmer world to live in.

Planes, trains, and automobiles

By the twentieth century, there were plenty of trains, but if you wanted to go somewhere that trains didn't, or had a lot of luggage, something else was needed . . . some kind of personal vehicle.

But steam engines were too big to be fiitted into such small vehicles, and they were also so heavy that there was no chance of a steam-powered aircraft. If you wanted to fly, and you weren't a bird, the only vehicle on offer was a hot air balloon—which, not being steerable, was more fun than useful.

A new kind of engine was invented called an internal combustion engine, powered by **burning fuels** made from oil. These new engines were light (and oil packs in more energy, weight for weight, than coal does) and (fairly) efficient. Cars, planes, and motorbikes could all be developed, and gradually trains and ships switched to using these new engines as well.

How an internal combustion engine works

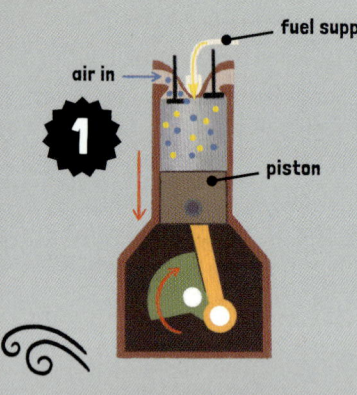

fuel supply

air in

1

piston

In an internal combustion engine, fuel is squirted into the cylinder while the piston is falling ('intake' phase), and air is sucked in.

2

The gas and air mixture is then compressed (made smaller) by the rising cylinder ('compression' phase).

The internal combustion engine took a long time, and a lot of inventors, to develop, but it was Karl Benz who, in 1879, invented a really good version. He went on (in 1886) to build cars that used the engine, though they didn't look much like today's models.

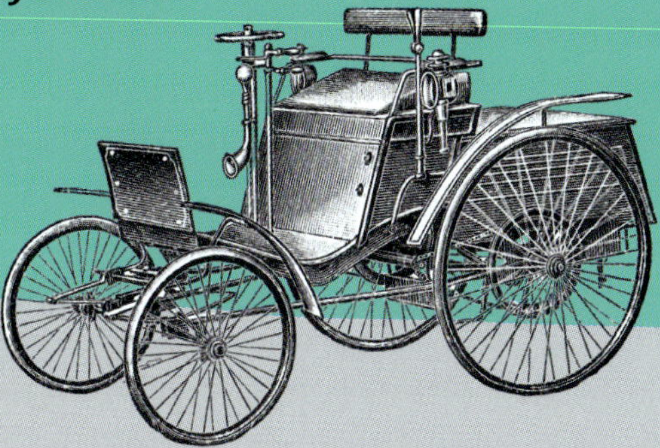

3

Then, an electric spark makes the fuel explode, pushing down the piston ('power' phase).

4

gases out

Finally, as the piston rises again, it pushes out the gases left over by the explosion ('exhaust' phase).

This four-stage cycle then begins again.

As well as coal and oil, a third fossil fuel called natural gas became widely used in the UK in the 1810s. Because it burned easily and brightly, and because it could be pumped along pipes, gas was ideal for lighting streets, factories, and homes, which until then had to rely on dingy oil lamps and feeble candles. Well, 'ideal' is a bit of an exaggeration—the gas did have a nasty habit of escaping and then poisoning you, not to mention **blowing UP** and/or burning down your house.

Both oil and natural gas were formed millions of years ago, from long-gone sea creatures and marine plants, most of which were too tiny for the human eye to see (not that there were any humans around to see them). When they died, they drifted down to the bottom of the sea and were slowly covered by thick layers of mud. The weight of the mud and heat from the Earth slowly changed them into oil and natural gas.

Unfortunately, burning all these fossil fuels turned out to have various **catastrophic effects** on the planet, as we'll discover in the final chapter.

Electrifying inventions

Although people had learned to change the chemical energy of coal and oil to kinetic energy by using engines, there was no way to move that energy around except by moving the fuel itself. Steam engines and ships had to drag tons of coal around with them, and people had to have coal and oil delivered to their homes.

All this was costly, time-consuming, dirty, and took up space. And when it came to vehicles, the weight of all that fuel meant more work for the engine, which meant even more fuel. (This is still a real problem for spacecraft, which need most of their fuel just to carry the rest of their fuel into space.)

But then Michael Faraday discovered how to change kinetic energy into electrical energy. The gadget that did the trick is called a dynamo (or generator). Like all great inventions, it's super simple—just spin a metal disc through the middle of a curved magnet and electrical energy will **flow down wires** attached to the disc.

Dynamos work because of the way electrical and magnetic energy work together. The magnet drags on electrons in the metal, making them move through the disc and along the wire. When they get to the bulb, some of their energy changes to thermal energy and to light.

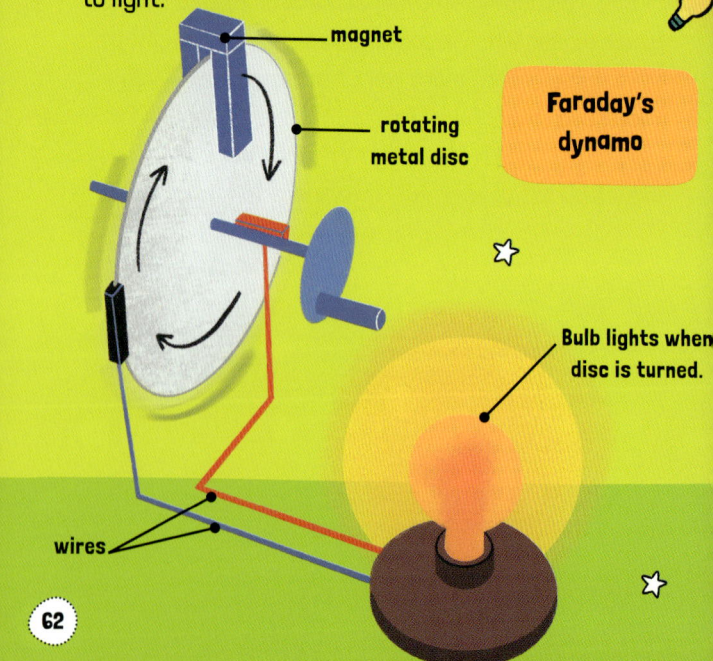

magnet

rotating
metal disc

Faraday's
dynamo

Bulb lights when
disc is turned.

wires

Today, dynamos have many designs—sometimes it's the magnet that moves—but all use the same idea of magnets dragging electrons in metals to make them move.

There are lots of ways to spin a dynamo:

- on some bikes, the pedals do the job and the electricity lights the bike lights

- on windfarms, it's **turbines** that spin in the wind

- in power stations, steam (from water boiled either by burning fossil fuel or by a nuclear **reactor**) is blasted on to turbines to do the same thing.

Thanks to the wonderful fact that electrical energy travels down wires, you can make your energy wherever you like, send it as far away as you can lay wires to, and just plug in your light bulb or supercomputer at the other end.

It isn't even necessary to plug in. Batteries contain chemicals which release electrical energy when needed. Rechargeable batteries can reverse this process, changing electrical energy into chemical energy.

People have been using most kinds of energy for **thousands of years.** But one form is much more recent, as we will now discover . . .

Chapter 5

Atomic World

The Industrial Revolution and the harnessing of electrical energy showed that science could be very useful, as well as great fun. Suddenly, it wasn't only scientists who loved it. Michael Faraday started to give lectures to the public, governments funded more schools, and more popular science books were written. Thanks to a very rich scientist called Alfred Nobel, even cash prizes and medals were now on offer! Soon, there were lots more scientists making lots more discoveries, including . . .

. . . Antoine-Henri Becquerel who, in 1896, discovered that a rare metal called uranium glowed with a previously unknown kind of radiant energy. Very few things—like water, air, and windows—let light pass through them (that is, they are transparent). But the new radiation could pass through almost **EVERYTHING**.

Soon, Marie Curie and her husband Pierre discovered two new materials (radium and polonium) which shone brightly with this **strange** energy. The radiation is now called gamma radiation (or gamma rays).

Speak like a scientist

RADIOACTIVITY

A great deal of energy is stored in the nuclei of atoms. When this nuclear energy is released, some of it appears as gamma rays and some as kinetic energy, carried by high-speed electrons and other particles. Together, these rays and particles are called radioactivity. Radioactivity is very dangerous—Marie Curie died as a result of her research, and even today her notebooks are too radioactive to handle safely.

ENERGY HEROES

MARIE AND PIERRE CURIE AND HENRI BECQUEREL

Unlocked the secrets of radioactivity and shared a Nobel Prize in 1903.

Nuclear fission

Centuries ago, alchemists tried to change lead into gold. But they couldn't do it, partly because they didn't understand what makes lead and gold different. In fact, the difference is in their nuclei.

Nuclei are like teeny bunches of grapes (called **nucleons**). The nucleus of a lead atom has 208 nucleons, eleven more than a gold nucleus.

So, to turn lead into gold, you just need to remove eleven nucleons from each nucleus. **But how?**

In 1931, John Cockcroft and Ernest Walton found out. But rather than changing lead into gold, they turned **lithium** (a metal) into **helium** (a gas).

How to make helium

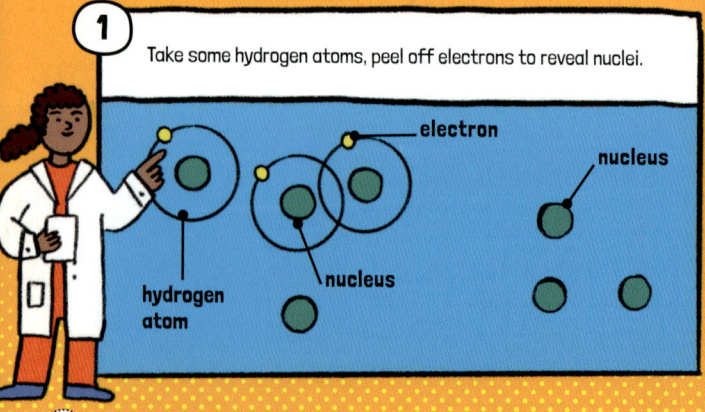

1 Take some hydrogen atoms, peel off electrons to reveal nuclei.

electron

nucleus

hydrogen atom

nucleus

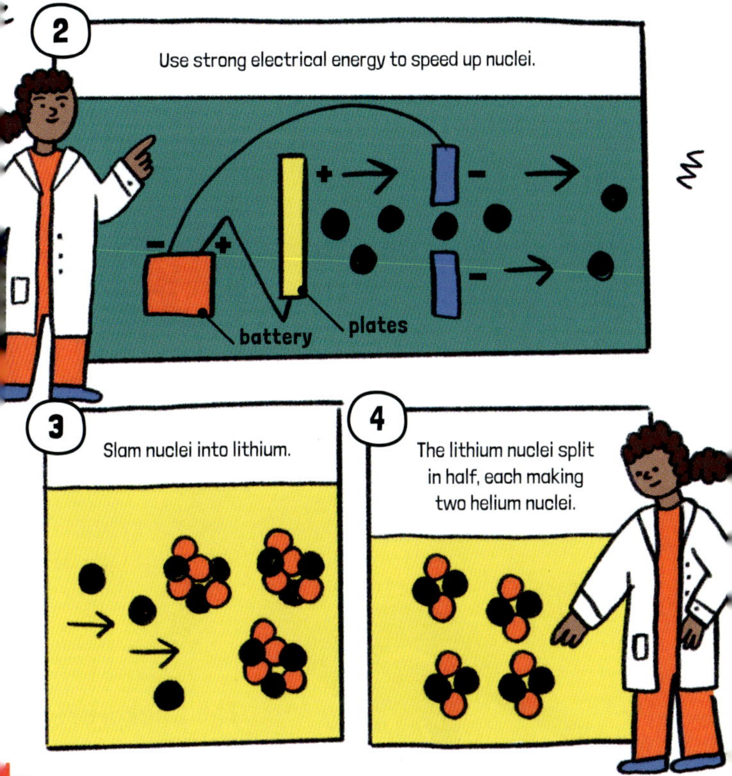

2 Use strong electrical energy to speed up nuclei.

battery · plates

3 Slam nuclei into lithium.

4 The lithium nuclei split in half, each making two helium nuclei.

'Splitting' is sometimes called 'fission', and today this process is called **nuclear fission.**

It's actually rubbish as a money-making scheme, since your electricity bill could never be covered by the money you might make selling your helium and/or gold. So, was nuclear fission a waste of time? And, by the way, why is this story in a book about energy anyway?
Well . . .

The black room

When Cockcroft and Walton weighed how much helium they had made from their lithium, they found a small difference: the mass was just a little less, because some had changed to kinetic energy, and then to heat, which could be very useful . . . Cockcroft and Walton got a Nobel Prize, and the world got a new source of power: nuclear energy.

Nuclear energy (occasionally called atomic energy) is similar to chemical energy—both are energy that is locked up inside atoms.

The Cockcroft Walton Generator

Cockcroft and Walton used this **generator** to provide the electrical energy they needed.

But whereas chemical energy is released when atoms move from one molecule to another (see page 31), nuclear energy is released when some of the nucleons (neutrons and protons) leave their nucleus and escape, or join up with others.

In the radioactive materials that Becquerel and the Curies had studied, the nuclei break down naturally. These breakdowns produce radioactivity, along with lots of energy and plenty of high-speed neutrons (a type of nucleon).

This led a number of scientists to an amazing idea: why not use the neutrons produced naturally from, say, uranium to **slam** into other uranium nuclei, to break them apart? This would release more neutrons, which would break more nuclei, and so on. If this 'chain reaction' could be controlled, it would mean a new source of power.

In 1940, a team of top scientists led by Enrico Fermi got to work to test the idea. The result was a huge **many-layered sandwich** (called a pile, or nuclear reactor) with thin layers of uranium and thick layers of graphite. The pile was a huge black, dust-covered block, built in a black squash court, by a team of dust-covered scientists.

Neutrons from uranium are very fast-moving and would simply escape from the reactor without breaking down any other nuclei, unless they were slowed down. The graphite (which is what makes your pencil black) was there to do this. A lot of it was needed: more than it would take to make a pencil for everyone on Earth.

**Fermi's first 'pile'
or nuclear reactor**

The reactor was incredibly dangerous—the uranium would change into **plutonium**, which is very radioactive and one of the most poisonous substances on Earth. Furthermore, the world's first atomic reactor could very easily turn into the world's first atomic bomb.

If the neutrons from each uranium nucleus caused two more nuclei to break down, and those two each broke down two more, and each of those . . . well, you can see where this is going. To control this (they hoped), the scientists used rods of a metal called cadmium, which captures neutrons.

In 1942, the reactor was ready. The cadmium rods were removed, and all eyes were on a measuring device. Sure enough, its needle quivered and rose, showing that radioactivity was being produced—the pile was working, and the world had entered its atomic age.

In 1945, two years after the reactor started working, **atomic bombs** were dropped by the USA on the Japanese cities of Hiroshima and Nagasaki, to end the Second World War. They killed over 100,000 people.

ENERGY HERO

ENRICO FERMI

Built the first nuclear reactor and discovered some of the secrets of atomic nuclei.

Harnessing nuclear energy

In 1951, the world's first experimental nuclear power station started producing electricity, which began to be supplied to homes and factories from 1954. Nuclear power stations are like fossil-fuel-burning ones, except that the heat to make the steam to turn the turbines is provided by a nuclear reactor.

nuclear-powered submarine

In the 1950s, there was a lot of excitement about nuclear power: nuclear-powered submarines, planes, and even cars and interstellar spacecraft were planned. But soon it became clear that there were all sorts of problems. Nuclear reactors produce **dangerous radioactive waste** that has to be buried deep in the Earth. Faults in power stations could lead to waste spreading by air and water, or even to devastating explosions. And once animals or people came into contact with the waste, they often became ill, either straight away, or years later through cancer. Today, the only nuclear-driven vehicles are submarines.

Pluto

Long-range space
probes are supplied
with electricity by
nuclear-electric generators.
These are thermocouples (see page 17)
heated by radioactive materials. The probes
use them to power their measuring equipment,
cameras, radio systems, and other devices,
not to move themselves through space.

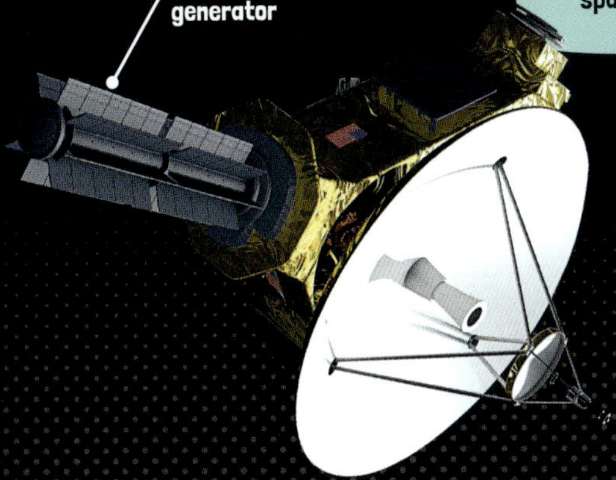

thermoelectric
generator

New Horizons
space probe

Fission and fusion

Today, about 10% of the world's electrical energy comes from nuclear power stations. But thanks to their dangerous waste and occasional horrific accidents, they are controversial.

Today's power stations use the fission (breaking apart) of the nuclei of massive atoms to get their electricity. The atoms are so big they're just itching to break apart anyway and so need little encouragement to do so (this is why their waste is so dangerous— their nuclei go on falling apart for thousands of years, constantly releasing radioactivity). But nuclear energy is also released when the nuclei of small atoms are **crushed** together. This happens in the Sun, where huge gravity forces press **hydrogen** nuclei so close that they 'fuse', sticking together to form the nuclei of helium and other types of atom. This is called **nuclear fusion** (which, since it sounds quite like fission, can lead to terrible con-fusion).

A fusion-based nuclear power station would have many advantages over the fission-based ones we use today: no need for dangerous fuels, no dangerous waste to get rid of. And there's no chance at all of a **nuclear explosion**: as soon as the crushing power is cut off, the fusion stops.

Using very strong magnets, we can just about **squeeze** atoms hard enough to make them fuse and release energy. But so much energy is needed to crush them together that it's not yet possible to build a reactor that supplies more electricity than it uses.

Experimental fusion tokomak

In a tokomak, a hot plasma is squeezed so tightly by magnetic fields that fusion takes place. To work well, the magnets must be kept very cold.

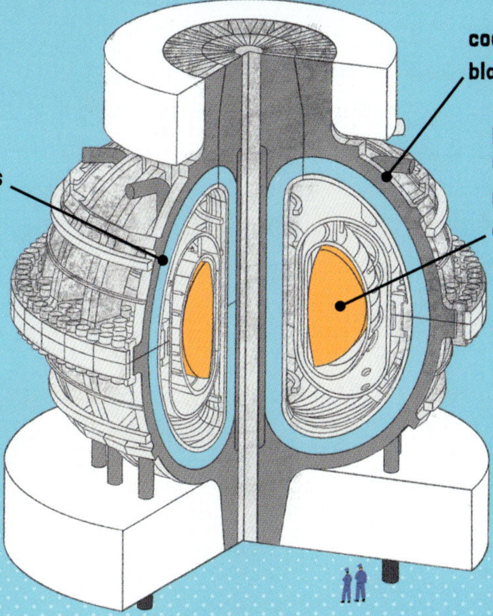

cooling blanket

plasma (hygrogen nuclei and electrons)

magnets

Energy is easy to free, once you know how. But freeing too much, or the wrong kind, can be a menace . . .

Chapter 6

Going Green

All the machines that surround us change energy from one form (or more) to one or more others. But, unless we build them very cleverly and use them very carefully, they can fill our world with waste. All machines waste some of the energy from their fuel as heat—and poorly designed ones waste a lot.

This was a worry during the Industrial Revolution because wasted energy meant extra fuel, and that meant higher costs. By the 1960s, it was clear that badly designed machines caused other problems too. The waste they produced included all kinds of dangerous chemicals, as well as unwanted sound energy (noise).

In the 1970s, laws were passed in many countries to try to stop the oceans, rivers, lands, and air from becoming **even more polluted.** In many cases, these were quite successful. But one waste product we can never get rid of is thermal energy. However well we design a machine, there is only one kind which is 100% efficient—that is, which converts all the energy we put into it into the form we want—and that is a heater.

It's just not possible to make a 100% efficient steam engine, car, refrigerator, or anything else. The best car (petrol) engines are about 20% efficient, and it's impossible to make one more than 37% efficient. The reason is that, however cool the exhaust gases might be, they still carry some of the engine's energy away with them in the form of heat.

The maximum efficiency here (like that 37%) is called the Carnot Efficiency, after the chap who explained it in 1824.

ENERGY HERO

NICHOLAS CARNOT

Showed how to calculate the maximum possible efficiency of any engine.

The greenhouse effect

Almost all the energy that we use comes from the Sun, including that stored as chemical energy in fossil fuels. Plants use solar energy **to grow**, which means our food supplies depend on the Sun too.

All planets in our Solar System receive energy from the Sun in the form of light and other kinds of radiant energy, and this increases their thermal energy. Some of this radiates away from the planet again, mostly as infrared. The temperature of the planet depends on the balance between the energy that arrives and the amount that leaves.

Some planets, including Earth, are surrounded by gas layers called atmospheres. These cut off some of the radiant energy from the Sun and also some of the infrared from the planet. But they are better at cutting off the infrared. So, they keep their planets warm, like blankets (this is called the greenhouse effect, even though greenhouses don't work this way). If the Earth had no atmosphere, it would be about 30°C colder.

The effect was discovered in 1859 by John Tyndall, who also discovered why the sky is blue (the explanation is tricky, but the main point is that blue and violet light from the Sun spreads through the atmosphere much more than the other colours in sunlight). You'd think he'd be more famous, really.

Some gases in the atmosphere, including carbon dioxide, cause a much stronger effect than others, so they are called greenhouse gases. Because the atmosphere of Venus is almost all carbon dioxide, it is about 500°C hotter than it would be with no atmosphere.

The amount of carbon dioxide in our atmosphere has been increased by the burning of fossil fuels—our world is now about **1°C hotter** than if the Industrial Revolution had never happened.

ENERGY HERO

JOHN TYNDALL

His experiments proved the greenhouse effect, and he showed why the sky is blue.

Global warming

The climate is changing, and a warmer world is bad news. The water in the ocean expands as it warms, causing sea levels to rise. **Melting** polar ice raises the sea levels further. The result is flooding, especially in areas where there isn't enough money to strengthen flood defences. Meanwhile, some supplies of drinking water dry up. Extreme weather events (heatwaves, flooding, heavy snow) and the loss of life and crops they cause, become much more likely and more extreme. Wildfires also become more frequent and more likely. Extreme weather events are thought to have become five times more common than they were fifty years ago.

When we burn fossil fuels, carbon dioxide is produced, and the weight of carbon dioxide that a fuel-burning factory or car produces is called its carbon footprint.

There are three main ways to reduce global warming:

1. Reduce energy use — perhaps by travelling less or getting a bus, cycling, or walking.

2. Absorb more carbon dioxide by planting new forests.

3. Use energy sources with smaller (or no) carbon footprints.

None of these ways of cutting back on global warming are easy. Cutting back on travel reduces people's choices of schools, jobs, and holidays. Not all governments take global warming seriously, and poorer countries have no choice but to use more energy if they are to get wealthier and improve the quality of the lives of their populations.

Today, the most popular way to tackle global warming is choosing **different energy sources**.

This group of energy sources has several names, which mean slightly different things:

Speak like a scientist

RENEWABLE

'Renewable' or 'sustainable' energy comes from a source which will not run out for many thousands of years, like the wind, the tides, the heat from inside the Earth, or sunlight.

GREEN

'Green' energy does not produce any pollution when it is made. Solar-power systems are green, for example (see page 85).

CLEAN

'Clean' energy is green energy produced in a way that causes no environmental problems. But no systems that release energy for us are completely clean: wind turbines can harm birds and cause noise pollution, hydroelectric systems can flood land previously used for farming, and any large new power station will have a big impact on the area where it is built.

LOW CARBON

'Low carbon' energy usually means any energy other than that from fossil fuels.

ALTERNATIVE ENERGY

'Alternative' energy is quite vague—it refers to any energy source that has not been used on a large scale for long, like wind farms. So, hydroelectric power stations are not 'alternative', but they are 'green' and 'renewable'.

Water power

The most popular green energy source is water. The first known watermills, which use the kinetic energy of **moving water** to grind corn into flour, were used in Greece over 2,000 years ago. Today, water supplies most of the world's green energy. Usually, this is by means of hydroelectric power systems.

More recently, the kinetic energy of waves and the potential energy of the tides—which are the twice-daily **rise** and **fall** of the world's oceans—have been used to produce electricity.

How hydroelectricity is made

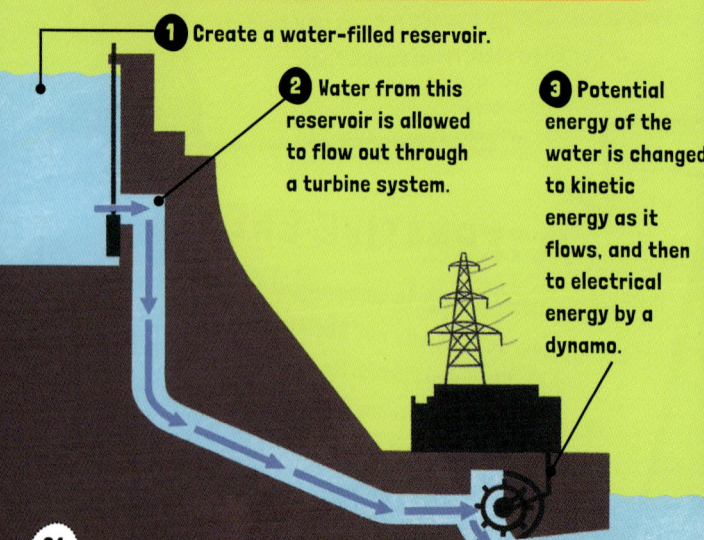

1 Create a water-filled reservoir.

2 Water from this reservoir is allowed to flow out through a turbine system.

3 Potential energy of the water is changed to kinetic energy as it flows, and then to electrical energy by a dynamo.

Wind and sun

There are many ways to capture the power of the Sun for our own use: laying water pipes on sunny rooftops can provide plenty of hot water easily. For cloudy days, these systems are combined with a gas or electric boiler. Or, we can use solar cells or panels, which are devices that convert the radiant energy of sunlight to electricity.

HOME SOLAR ENERGY SYSTEM

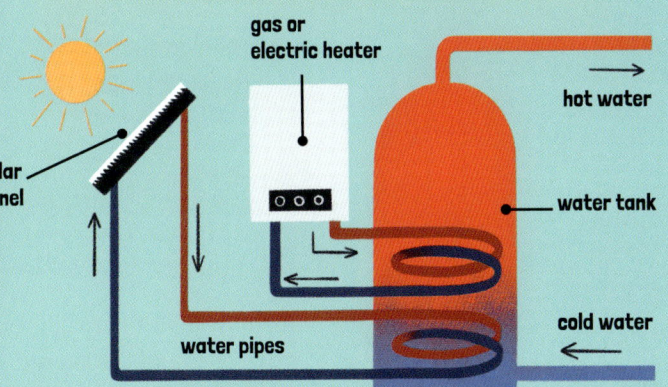

gas or electric heater

hot water

solar panel

water tank

water pipes

cold water

A big advantage of solar power systems is that people can install them in their homes, so there's no need for long power cables from power stations. These cables are expensive to make, lay, and repair, often ugly, and they waste some of the electricity they carry too.

The Sun makes the wind, by heating some parts of the Earth more than others. The air there expands, becomes lighter and **rises.** The air from nearby cooler areas rushes in, and this rushing is the wind. Our ancestors used the wind to blow their ships across the oceans, and to turn windmills to grind flour.

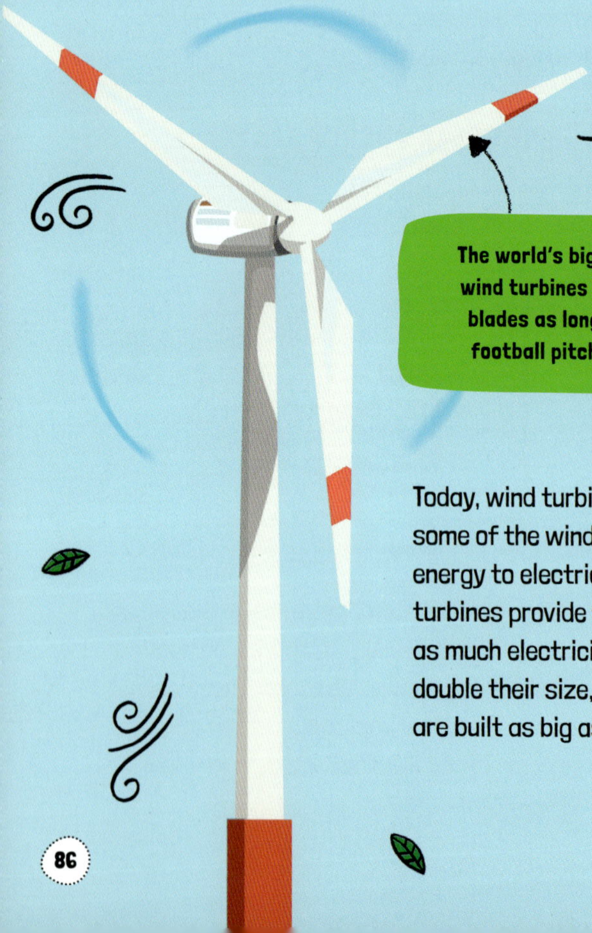

The world's biggest wind turbines have blades as long as football pitches.

Today, wind turbines change some of the wind's kinetic energy to electricity. These turbines provide four times as much electricity if you double their size, so they are built as big as possible.

Heat from the deep

The Earth formed over **four billion** years ago, when gravity pulled together a vast number of rocky and metallic lumps into one giant sphere. The kinetic energy of all these fast-moving lumps changed to thermal energy when they collided, so much so that the whole new planet Earth melted.

The outer layers of the Earth cooled fast and were soon solid rock again, but the deep interior is still very hot today—in the centre, Earth's temperature is higher than that of the surface of the Sun.

Living on such a hot–centred world is both good news and bad. The bad news is that the Earth's underground thermal energy causes **earthquakes and volcanoes.** The good news is that we can use it. In Iceland, water warmed by hot rock is used to heat some houses, while in Japanese winters, monkeys laze around in it.

A geothermal system

Usually, the Earth's heat is not so easy to get at. One way is to drill two holes down to some nice hot rocks and pump water down one. The rocks boil the water into very hot steam, which rushes up the other hole, and a turbine and generator change its kinetic energy to electricity. The hot steam can also be used for home-heating.

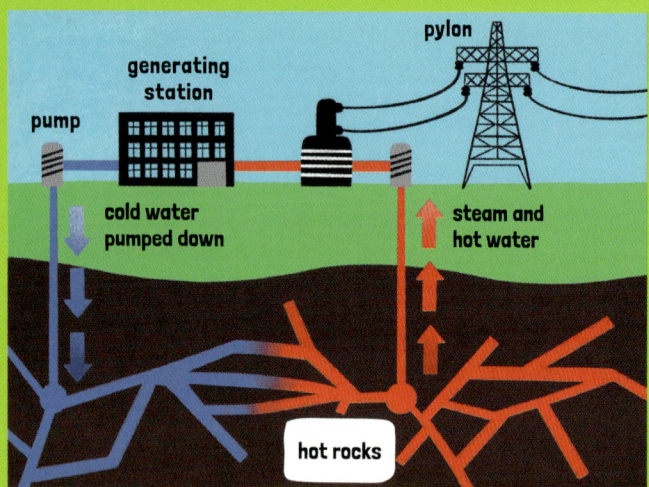

pylon

generating station

pump

cold water pumped down

steam and hot water

hot rocks

Burning something better

But there are other things that burn: hydrogen gas, for example, and all this produces (other than thermal and radiant energy) is water.

Another advantage of using hydrogen as a fuel is that we don't need to pipe it to power stations to turn it into electricity. A device called a fuel cell, small enough to put in a cupboard or a car, allows the hydrogen to combine with oxygen to produce electrical energy, plus water.

The problem is, we can't just find hydrogen in the Earth like we can fossil fuels, we have to make it. The simplest way is by using electrical energy to break down water into hydrogen and oxygen (this is called electrolysis). But the process is expensive and of course it's pointless unless the electricity comes from a low carbon source.

Another way to get hydrogen is to break down methane gas using **hot steam**. This does produce some carbon dioxide, but it is a cheaper process.

A different approach is to break down plant waste to make a chemical called ethanol, and then burn that. A similar process breaks down waste oil to make a fuel called biodiesel. Ethanol and biodiesel are biofuels, which are fuels made from animal or vegetable matter, but made recently, not millions of years ago like fossil fuels. However, the amount of carbon dioxide biofuels produce when they are burned is still more than half as much as would be produced by burning fossil fuels.

Now and tomorrow

Energy was a mystery just a few centuries ago, but today, thanks to the work of scientists, engineers, and inventors, we have learned not only what it is, but how to control it.

You are surrounded by energy-control devices right now—from phones to computers and from central heating systems to lightbulbs. The foods you eat, the ways you travel, and the clothes you wear are all the result of **energy control**.

But our convenient modern world has come at a cost. Global warming and pollution are problems for us all. Although people are trying to tackle these challenges all over the world, our planet has already been changed, and we need to cope with living in a warmer, wetter, and more unpredictable world.

Green energy cannot be easily stored, and can only be made in certain locations, or when the weather is right.

At the moment, all the ways we get energy have their problems. Releasing chemical energy from fossil fuels causes global warming and pollution. Nuclear fission is dangerous and polluting.

Science will solve some of these problems. If we can make fusion reactors which make more electrical energy than they use, they could replace all other power stations. Batteries that store much more energy would make green energy easier to use. And science can tell us a lot about what any new device will give us: how much useful energy, but also how much waste and pollution. But science has its limits: it can't tell us how to use energy in a completely safe and clean way.

Instead, we must make careful choices, and difficult decisions.

Energy has given us
incredible powers.
Now we must learn to use them wisely.

Glossary

atom a tiny lump of matter, much too small to see, from which every object is made

carbon dioxide a gas produced by burning and breathing

charge the amount of electricity in something

chemical a substance obtained by or used in chemistry

dynamo a machine which changes kinetic energy to electrical energy

efficiency the amount of useful energy a device produces compared to the amount it uses

electrons tiny particles which form the outer parts of atoms and which carry electrical energy

force a push or pull

fossil fuel coal, oil, or burnable gas from underground

helium a light colourless gas that does not burn and is sometimes used to fill balloons

hydroelectric using water power to produce electricity

hydrogen a lightweight gas that combines with oxygen to form water

infrared a kind of radiation sent out by warm objects

joule measure of energy, abbreviated as 'J'

lithium the lightest metal in the Universe

mass a form of energy which gives things their weight

matter anything that is not energy

molecule two or more atoms joined together

motor a machine that changes electrical or chemical energy to kinetic energy

neutron a particle usually found in the centre of an atom, but which can be released when atoms break apart

nuclear fission the breaking apart of the nuclei of atoms. The energy released is used in some power stations and weapons

nuclear fusion the joining together of the nuclei of atoms. The energy released makes the Sun shine

nucleons particles that are joined together to form the nucleus of an atom—for example, neutrons are one kind of nucleon

nucleus the centre of an atom (plural: nuclei)

oxygen a colourless odourless tasteless gas that exists in the air and is essential for living things

piston a disc or cylinder that fits inside a tube in which it moves up and down as part of an engine or pump

plutonium a dangerously radioactive metal

radiation light and other energy of the same kind, including infrared and gamma rays

radioactivity fast particles and radiation released by materials in which nuclear fission is happening

reactor device in which nuclear energy is released in a controlled way

trillion a million million

turbine fan-like device which is turned by wind, water, or steam; usually connected to a dynamo to change kinetic energy into electrical energy

ultraviolet a kind of radiation which darkens human skin and can cause cancer

unit an amount used as a standard in measuring or counting things

uranium radioactive metal used as a nuclear fuel.

watt a measure of power, abbreviated as 'W'

Index